YOUR KNOWLEDGE HAS VALUE

- We will publish your bachelor's and
  master's thesis, essays and papers

- Your own eBook and book -
  sold worldwide in all relevant shops

- Earn money with each sale

Upload your text at www.GRIN.com
and publish for free

Anti-fungal activity of traditional spices against dermatophytes and opportunistic fungi. Comparison of combinatorial effects of clove, cinnamon and kacholam

R. Mohanan

S. Thomas

S. P. Jose

S. Sreevallabhan

S. Sukumaran

G. Bhaskaran Nair

A. Sukumarapillai

S. Rajan

J. Joseph

Bibliographic information published by the German National Library:

The German National Library lists this publication in the National Bibliography; detailed bibliographic data are available on the Internet at http://dnb.dnb.de.

ISBN: 9783346296849
This book is also available as an ebook.

© GRIN Publishing GmbH
Nymphenburger Straße 86
80636 München

All rights reserved

Print and binding: Books on Demand GmbH, Norderstedt, Germany
Printed on acid-free paper from responsible sources.

The present work has been carefully prepared. Nevertheless, authors and publishers do not incur liability for the correctness of information, notes, links and advice as well as any printing errors.

GRIN web shop: https://www.grin.com/document/953223

ANTI-FUNGAL ACTIVITY OF TRADITIONAL SPICES AGAINST DERMATOPHYTES AND OPPORTUNISTIC FUNGI: COMPARISON OF COMBINATORIAL EFFECTS OF CLOVE, CINNAMON AND KACHOLAM

1.   **Dr. Ratheesh Mohanan (Corresponding author)**
2.   Ms. Sulumol Thomas
3.   Ms. Svenia P Jose
4.   Ms. Sheethal Sreevallabhan
5.   Dr. Sandya Sukumaran
6.   Dr. Girishkumar Bhaskaran Nair
7.   Ms. Asha Sukumarapillai
8.   Mr. Sony Rajan
9.   Mr. Jobin Joseph

Mahatma Gandhi University, Kottayam

1- Department of Biochemistry, St. Thomas College, Palai, Kottayam, Kerala, India

Contents

List of Figures..iii

List of Tables..V

Abbreviations iii

Abstract 1

1. Introduction..2

    1.1. Objective of the study 3

2. Review of literature 4

    2.1. Spices...4

    2.3. Opportunistic Fungi- *C. albicans*..8

3. Materials and Methods 9

    3.1 Preparation of Raw material 9

    3.2 Extraction process 9

    3.3 Preparation of Media 10

    3.4 Fungal strains and cultural conditions 10

    3.5 Haemocytometer counts...10

    3.6 Anti-fungal screening of extracts 11

4. Results 11

    4.1 Screening of DMSO (Negative control) and standard drugs (Positive control)..........................11

    4.2 Determination of antifungal effects of clove, cinnamon, kacholam against *T. mentagrophytes*12

    4.3 Determination of antifungal effects of clove, cinnamon, kacholam against *T. rubrum* 14

    4.4 Determination of antifungal effects of clove, cinnamon, kacholam against *C. albicans* 16

    4.5 Determination of combinatorial effects of clove, cinnamon, kacholam against pathogenic fungi 18

5. Discussion 20

6. Summary and Conclusion 22

References...27

List of Figures

Figure No.	Title	Page No.
2.1	General Classification of Dermatophytes	5
2.2	Culture of *T. rubrum* and its microscopic view	7
2.3	Culture of *T. mentagrophytes* and its microscopic view	7
2.4	Culture of *C. albicans* and its microscopic view	8
4.1	a) Zones of inhibition produced by standard drug Flc and Ket against *T. mentagrophytes*. b) Zones of inhibition produced by standard drug Flc against *C. albicans*. c) Zones of inhibition produced by standard drug Ket against *T. rubrum*.	11
4.2 i)	Graphical representation of inhibitory effects of clove, cinnamon and kacholam against *T. mentagrophytes*	13
4.2 ii)	a. Anti-fungal effect of kacholam against *T. mentagrophytes*. b. Anti-fungal effect of clove against *T. mentagrophytes*. c. Anti-fungal effect of cinnamon against *T. mentagrophytes*.	14
4.3 i)	Graphical representation of inhibitory effects clove, cinnamon and kacholam against *T. rubrum*.	15
4.3 ii)	a. Anti-Fungal effect of clove against *T. rubrum*. b. Anti-Fungal effect of cinnamon against *T. rubrum*. c. Anti-Fungal effect of kacholam against *T. rubrum*.	15
4.4 i)	Graphical representation of inhibitory effects clove, cinnamon and kacholam against *C. albicans*.	17
4.4 ii)	a. Anti-fungal effect of clove against *C. albicans*. b. Anti-Fungal effect of cinnamon against *C. albicans*. c. Anti-Fungal effect of kacholam against *C. albicans*.	17
4.5 i)	Graphical representation of inhibitory effects of combined extracts against pathogenic fungi.	19

4.5 ii)	a. Anti-fungal effect of clove-cinnamon-kacholam against *T. rubrum*. b. Anti-fungal effect of clove-cinnamon-kacholam against *T. mentagrophytes*. c. Anti-fungal effect of clove-cinnamon-kacholam against *C. albicans*.	19
6.1	Schematic representation of antifungal effects of clove, cinnamon and kacholam against *Trichophyton rubrum*.	22
6.2	Schematic representation of antifungal effects of clove, cinnamon and kacholam against *Trichophyton mentagrophytes*.	23
6.3	Schematic representation of antifungal effects of clove, cinnamon and kacholam against *Candida albicans*.	24
6.4	Schematic representation of combined effect of clove, cinnamon and kacholam against selected pathogenic fungi.	25

List of Tables

Table No	Title	Page No.
4.2	Inhibitory effects clove, cinnamon and kacholam against *T. mentagrophytes*	12
4.3	Inhibitory effects clove, cinnamon and kacholam against *T. rubrum*	14
4.4	Inhibitory effects clove, cinnamon and kacholam against *C. albicans*	16
4.5	Inhibitory effects of combined extracts against pathogenic fungi	18

Abbreviations

CDC	Control and Prevention
CFU	Colony Forming Unit
DMSO	Dimethyl sulphoxide
DNA	Deoxyribonucleic acid
Flc	Fluconazole
Ket	Ketoconazole
mg	Milligram
OI	Opportunistic infections
R	Resistance
RNA	Ribonucleic acid
S	Sensitive
Spp	Species
SDA	Sabouraud Dextrose Agar

Abstract

Spices have been used as food and flavouring agent since the ancient times, and as a medicine in the recent decades. Now it is widely used across the globe as they possess great potential in the treatment of various diseases. It can serve as a better alternative to the modern synthetic drugs due to its lack of side effects and possible role in a wide range of therapeutic application. The present study aims to determine the anti-fungal activity of combined and individual effect of Cinnamon *(Cinnamomum verum)*, Clove *(Syzygium aromaticum)* and *Kacholam (Kaempferia galangal)* against *Candida albicans, Trichophyton rubrum,* and *Trichophyton mentagrophytes* by agar well disc diffusion method. Extract from these spices showed remarkable antifungal activity against *T. mentagrophytes, T. rubrum* and *C. albicans.* Clove exhibits significant antifungal activity against all the microorganisms tested. Cinnamon showed good inhibitory effect against *T. mentagrophytes* and *C. albicans* whereas kacholam showed an inhibitory effect against *T. mentagrophytes* and *T. rubrum.* Thus, the study concluded that individual effect of spices is better than the combinatorial effect against all the fungus tested.

Keywords: Spices; Dermatophytes; Opportunistic fungi; Agar well diffusion method

1. Introduction

The history of medicinal plants is similar to that of the advancement of human history. Since ancient times, medicinal plants have been found to be the strength and backbone of traditional herbal medicine across the worldwide. Natural products are the fundamental part of different ancient medicinal systems like Ayurveda, Chinese, Greek- Unani (Sarker & Nahar, 2007). It is estimated that 40% of the world's population depending directly on plant-based medicine for their health care. India has a rich medicinal plant flora consist of more than 25,000 species, of which 150 are commercially used for the extraction of medicines or for the formulation of drugs (Lucy & Edgar, 1999).

Over the past decades, researchers have sought to identify and validate plant-derived substances for the treatment of various diseases. Natural medicine comprises of spices, herbs and shrubs. The spices and herbs have been considered as a part of the food in daily life. In the dietary system of humans, the addition of spices is essential for the enhancement of the taste and aroma of food items. Besides being used as flavouring agent, it is also having lots of therapeutical properties such as antioxidant, antimicrobial and anti-inflammatory activity. Spices are one of the common active ingredient used in traditional system of medicine or naturopathy because of its effectiveness against certain diseases (Langner et al, 1998). Moreover, Essential oils and extracts derived from spices are capable of controlling microorganisms related to skin diseases, dental caries and food spoilage. (Chaieb et al, 2007). A number of studies have been conducted in recent years on the use of spiceuticals as antimicrobial agents. In most cases, reports on the antifungal activity of some extracts or essential oils exposed directly to fungi have been also published (Baratta et al, 1998). Therefore, the use of spices could be an effective mode of treatment against superficial mycosis.

Superficial mycosis or cutaneous mycosis is a chronic disease condition which is commonly caused by certain dermatophytes. In human beings, *T. mentagrophytes* and *T. rubrum* are the most common pathogens which is causing skin infections. Dermatophytosis leads to the formation of cutaneous lesions which is characterized by desquamation, erythema of the edges and circular disposition (Hainer, 2003). Along with these dermatophytes, *C. albicans* an opportunistic fungus which can also cause mucosal and cutaneous infections along with subcutaneous infectious diseases and systemic mycoses. The previous studies showed that the active compounds present in certain spiceuticals or their extracts exhibit fungicidal activity without affecting the host.

The present study is focusing on the medicinal properties of traditional spices such as clove, kacholam and cinnamon against *T. mentagrophytes*, *T. rubrum* and *C. albicans*. A comparative analysis of the antifungal properties of these spices with that of different antibiotics will provide a valuable information regarding the antifungal activity of spices and its therapeutic potential targeting the replacement of common conventional drugs.

1.1. Objective of the study

Fungal infections are one of the common type of infection, but it can cause many serious clinical manifestations. Opportunistic infections (OIs) are the main leading causes of morbidity and mortality in immune-compromised patients. The discovery of antifungal agents was a major milestone in tackling the fungal infections, either as a cure or to manage the symptoms. But, the prolonged use of these agents can cause various side effects. Moreover, the human dermatophytic fungus will acquire resistance and tolerance against these antifungal agents. Recently, the opportunistic fungi have shown azole resistance and this might be the reason for the drastic emergence of Non- *C. albicans* species, as a cause of refractory mucosal candidiasis (Rex et al, 1995; Martins et al, 1997). Thus, in search of new and better alternatives, natural products are found to be very efficient because they have no detrimental effects on human beings but it could be deleterious to the microbial pathogens.

Therefore, the aim of the present study:

- To determine the anti-fungal effect of Clove, Cinnamon and Kacholam against *Trichophyton mentagrophytes, Trichophyton rubrum and Candida albicans.*

- To evaluate the combinatorial effect of Clove, Cinnamon and Kacholam, against *Trichophyton rubrum, Trichophyton rubrum and Candida albicans*

2. Review of literature

2.1. Spices

The Spice is a seed, fruit, root, bark or other plant substance primarily used to flavor, color or preserve food items. Spice is defined as a "strongly flavored or fragrant substance of natural origin, derived from tropical plants, frequently used it as a condiment".

International Organization for Standardization (ISO), defines the term Spices and Condiments are natural or vegetable ingredients or mixtures of the same, in whole or ground form, used for flavoring, flavoring and spicing of fruit and for seasoning of fruit. Spices are also considered to have many positive effects. The researchers shown that widely used herbs and spices such as garlic, black cumin, cloves, cinnamon, ginger, thyme, all-spice, bay leaves, mustard and rosemary have antimicrobial properties that can be used therapeutically in certain situations.

Clove is one of the most valuable spices that has been used as food preservatives for many centuries and has been used for many medicinal purposes and it is known as 'champion spice'. Cloves are native to Indonesia but are now cultivated in different parts of the world (Cortés-Rojas et al, 2014). Clove has many medicinal properties, including antiviral, antimicrobial, general antifungal stimulation, hypertensive aphrodisiac, light stomach, carmin and anesthetic. (Di Paoli et al, 2007).

Clove is known to have antibacterial properties and is used for cleansing bacteria in various dental creams, tooth pastes, mouth washings, and throat sprays. This is also used to relieve chronic gum pain and to improve overall dental health. Clove is an anodyne (an agent that relieves or soothes pain) in dental emergencies. It is used as an anti-inflammatory agent, because of its high flavonoids content. Clove and clove oil has the ability to strengthen the immune system by purifying the blood and helping to combat various diseases (Parle & Khanna, 2011).

Kacholam is a small monocotyledonous herb of Zingiberaceae that has been known for decades for its medicinal properties. The plant is native to southern China, Indochina, Thailand, Taiwan, Malaysia and India. Phytochemical isolated from kacholam, ethyl-p-methoxycinnamate has significant activity against *Mycobactrium tuberculosis* and *Candida albicans*. More recently, the resazurin microtitre assay has shown that this ethyl-p-methoxycinnamate will inhibit drug susceptible and multidrug resistant clinical isolates of *M. tuberculosis*. Kacholam extracts have also been found to exhibit antimicrobial activity against a number of organisms including both Gram positive and Gram negative organism.

Cinnamon is an important spice and aromatic crop with a wide range of applications in flavorings, perfumes, beverages and medicines. Cinnamon, also known as true cinnamon or ceylon cinnamon, is an evergreen tree with a strong aromatic bark and leaves.

Cinnamon is Native to Sri Lanka the largest producer and exporters. As well as being used as a spice and flavouring agent, cinnamon is also adding to the flavor of chewing gum because of its refreshing effects on the mouth and the ability to remove bad breath (Jakhetia et al, 2010). It improves colon health and thus reduces the risk of colon cancer (Wondrak et al, 2010). Cinnamon increases uterine blood circulation and promotes tissue regeneration (Minich et al, 2008). Cinnamon plays a vital role as a spice, but it also has significant activities in its essential oils and other constituents, including antimicrobial, antifungal, antioxidant and antidiabetic (Chang et al, 2001).

2.2.1 Dermatophytes

In general, dermatophytes are classified in three anamorphic genera, Epidermophyton, Microsporum, and Trichophyton belonging to the category of the Deuteromycota (Fungi Imperfecti) (Irene et al, 1995). This classification is based morphology of conidia and the conidial spore formation. Dermatophytes are among the few fungi that cause transmissible diseases, that is, diseases acquired from infected animals or birds or from the fomites they have endangered.

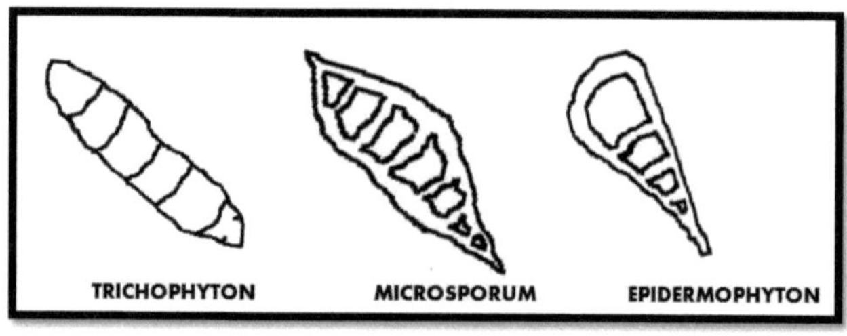

Figure 2.1. General Classifications of Dermatophytes[1]

---

[1] Author's own creation

Dermatophytes are aerobic fungi which will invade and infect keratinized layers of the skin, hair, and nails, mediated by both keratinases and proteases.

2.2.2 Epidermophyton spp.

Epidermophyton spp. have macroconidia which is narrowly clavate with usually smooth, thin to moderately thick walls and one to nine septa, 20 to 60 by 4 to 13 mm. They commonly abundant and are borne individually or in clusters. Microconidia are absent. To date, this genus includes only two known species, and only *Epidermophyton floccosum* can be pathogenic.

2.2.3 Microsporum spp.

Macroconidia of Microsporum spp. is characterized by rough walls that may be asperulated, echinulated, or verrucose. Emmons originally described macroconidia as spindle shaped or fusiform but the discovery of new species extended the range from obovate (egg shaped) as in *Microsporum nanum* (Fuentes et al, 1956) to cylindrofusiform as in *Microsporum vanbreuseghemii* (Georg et al, 1962). Microconidia are sessile or stalked and clavate and usually arranged singly along the hyphae or in racemes as in Microsporum racemosum, a rare pathogen (Borelli et al, 1965). The type species is *M. audouinii*.

2.2.4 Trichophyton spp.

In Trichophyton spp., macroconidia has smooth, usually thin walls and one to twelve septa, are borne individually or in clusters, and may be elongated and shaped like a pencil, clavate, fusiform or cylindrical. Microconidia, typically more numerous than macroconidia which can be globose, pyriform or clavate, or sessile or stalked, borne alone on the hyphae sides or in grape-like clusters. The type species is *T. tonsurans*.

2.2.5 *Trichophyton rubrum*

T. rubrum is an anthropophilic fungus that has become the most widely distributed dermatophyte on humans. It frequently causes chronic infections on skin, nails and infrequently on scalp. Granulomatous lesions may sometimes occur. Infected hairs will not fluoresce under Wood's ultraviolet technique, and it may show endothrix or ectothrix sort of invasion.

Approximately 80–93% of chronic dermatophytic infections reported in many parts of the world are thought to be caused by *T. rubrum* including cases of tinea pedis, tinea unguium, tinea manuum, tinea cruris, and tinea corporis, as well as some cases of tinea barbae. It has also been known to cause folliculitis which is characterized by fungal element in follicles and giant cells in the dermis. *T. rubrum* infection may also form a granuloma. Extensive granuloma formations may occur in patients with immune deficiencies (e.g. Cushing syndrome). Immunodeficient neonates are highly susceptible to systemic *T. rubrum* infection.

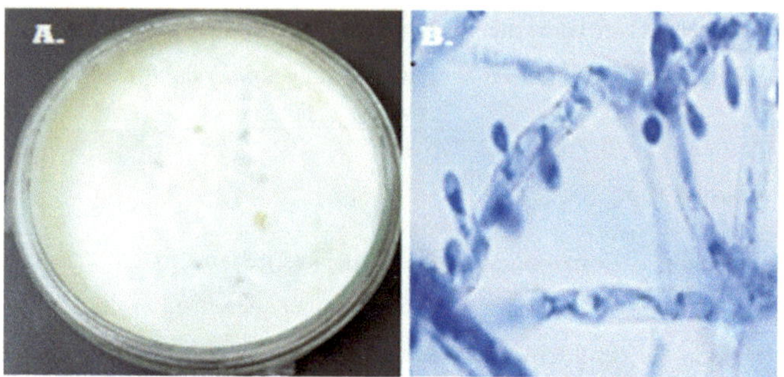

Figure 2.2. Culture of *T. rubrum* and its microscopic view[2]

2.2.6 Trichophyton mentagrophytes

T. mentagrophytes is a fungus that is part of a group known as dermatophytes. This fungus is known to cause skin infection known as Dermatophytosis or Ringworm which appears on a person's skin as an inflamed circular pattern. The invasion of this organism to skin, hair, and nails can cause diseases such as tines pedis or athlete's foot. Spores produced by this fungus are found to be difficult to remove by disinfection. Disinfection is especially important for environments where Ringworm infections can occur and spread rapidly such as athletic facilities or schools. *T. mentagrophytes* is zoonotic, meaning that it can be transferred from animals to humans (Martins et al, 2014).

---

[2] Author's own creation

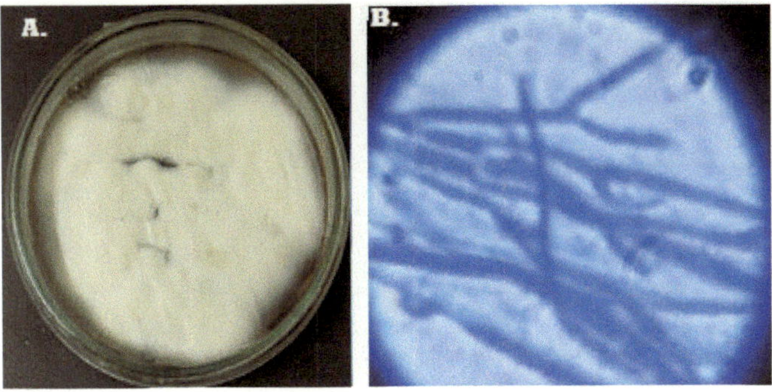

Figure 2.3. Culture of *T. mentagrophytes* and its microscopic view[3]

2.3. Opportunistic Fungi- *C. albicans*

Opportunistic fungi can be defined as fungi that would not normally cause infections in otherwise healthy people but they are capable of causing infection under certain circumstances, such as immunodeficiency, cancer, organ transplantation, neutropenic patients, diabetes, weakened patients and long-term antibiotic patients. It is generally a harmless commensal fungus that can turn into an opportunistic organism in immunocompromised or immunologically deficient individuals as in HIV patients.

C. albicans is a member of the human microflora as diploid polymorphic yeast of mucosal surfaces and is commonly found in the human gastrointestinal (GI), respiratory, and genitourinary tracts. *C. albicans* and other emerging NAC species, including *C. glabrata, C. krusei, C. tropicalis*, and *C. parapsilosis* represent an important source of systemic infections worldwide. These organisms are the most common cause of superficial vaginal or mucosal oral infections and may also, under propitious conditions, enter the bloodstream leading to deep-tissue infections (Conti et al, 2014). With a variety of host cells, including Th17 cell, this microorganism can interact during the disease manifestations (Kashem et al, 2015). An abnormal growth in *C. albicans* due to environmental imbalance like reduction of pH may result in candidiasis. It has been demonstrated that *C. albicans* systemic infections lead to a mortality rate of ~40%.

---

[3] Author's own creation

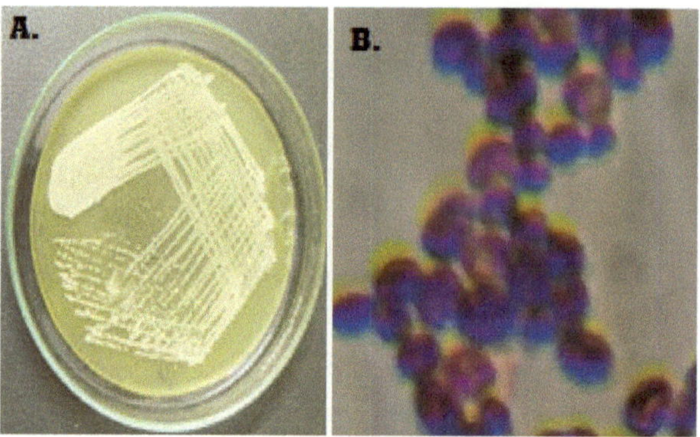

Figure 2.4. Culture of *C. albicans* and its microscopic view[4]

3. Materials and Methods

3.1 Preparation of Raw material

Dried clove, cinnamon and kacholam were purchased from local shop in Idukki. The samples were air-dried, grinded in a Wiley Mill to fine uniform texture, and stored in glass jars until use.

3.2 Extraction process

100g of powdered samples was taken in a conical flask and added required volume of 70% ethanol. The mouth of the conical flask was covered with aluminum foil and kept in a reciprocating shaker for 24 h for continuous agitation at 150 rev/min for thorough mixing and also complete elucidation of active materials to dissolve in the respective solvent. Then, extract was filtered by using muslin cloth followed by Whatman no 1 filter paper. The solvent from the extract was removed by using rotary vacuum evaporator with the water bath temperature of 60°C. Finally, the residues were collected and refrigerated at 4 °C until further analysis.

3.3 Preparation of Media

All ingredients/media were purchased from M/S Hi-Media Laboratories Pvt. Ltd. Mumbai, India.

- Sabouraud Dextrose Agar
- Mycological Peptone - 10 gms
- Dextrose - 40 gms .
- Agar agar - 15 gms
- Distilled water - 1000 ml
- pH - 5.4

Medium was mixed thoroughly and heated with frequent agitation. Boiled for 1 min and sterilized by autoclaving at 15 lbs/in' pressure for 15 min. About 25 ml medium poured into 90 mm diameter sterile glass plates to a depth of 4 ± 0.5 mm. Pouring was done on a level surface to obtain uniform depth of the medium.

3.4 Fungal strains and cultural conditions

All fungal strains used in this study including *C. albicans*, *T. rubrum* and *T. mentagrophytes* were procured from microbial culture lab. Fungal strains were cultured on Sabouraud Dextrose *Agar (HiMedia Laboratories) slants for 7-14 days at 28⁰C.*

Prepare a spore suspension by pouring 5–7 mL of sterile 0.9 % NaCl (w/v) + 0.01 % Tween 80 on a SDA plate containing a premade fungal culture. It was then scrapped by using sterol glass rod for about 2 min. It was then filtered using sterile cheese cloth and filtrate was used as stock for spore solution. A diminutive amount of the stock was diluted serially and spore counts were taken using haemocytometer (*Neubauer chamber, Germany*).

3.5 Haemocytometer counts

The haemocytometer and the coverslip were cleaned thoroughly in soap water followed by tap water. It was then rinsed in alcohol and wiped with a tissue paper. The cover slip was placed on the slide exactly over the depression in the counting chamber. One ml of spore suspension was drawn using the pipette and expelled into the depression below the cover slip.

After 3-5 minutes, spore count was taken under phase at 300-400 times magnification. For each dilution three replicates were maintained.

The number of spore ml -1 was calculated using the following formula

$$= n*10^4$$

n = Total number of spores

These plates were incubated for 24-72 hours at 25°C. The diameter of the zones of inhibition around each of the disc was taken as measure of the anti-fungal activity. Each experiment was carried out in triplicate and mean diameter of the inhibition zone was measured in millimeter (Kohner et al, 1994; Mathabe et al, 2006).

3.6 Anti-fungal screening of extracts

In vitro antifungal and anti-bacterial activity were determined by agar well diffusion method (Murray et al, 1995). For susceptibility testing, ethanolic extracts of spice were made into a suspension using DMSO (Merck). Fungal isolates were grown on SDA (HiMedia Laboratories). Extracts of each spices at different doses (6.25mg, 12.5mg, 18.75mg and 25mg) were introduced into the medium along with standard drug of ketoconazole (KT 50 mcg, HiMedia) and fluconazole (FLC 10 mcg, HiMedia). Then, a hole with a diameter of 6 to 8 mm is punched aseptically with a sterile cork borer or a tip, and a volume (50–200 μL) of the antifungal agent or extract solution at desired concentration is introduced into the well.

4. Results

4.1 Screening of DMSO (Negative control) and standard drugs (Positive control)

For the present study, negative and positive controls were examined for their antifungal activity against three fungal isolates by the agar well diffusion method. These are the fungal isolates were frequently encountered with human infections. The negative control DMSO 10% did not show any antifungal activity while the positive control KT, FLC showed antifungal activity.

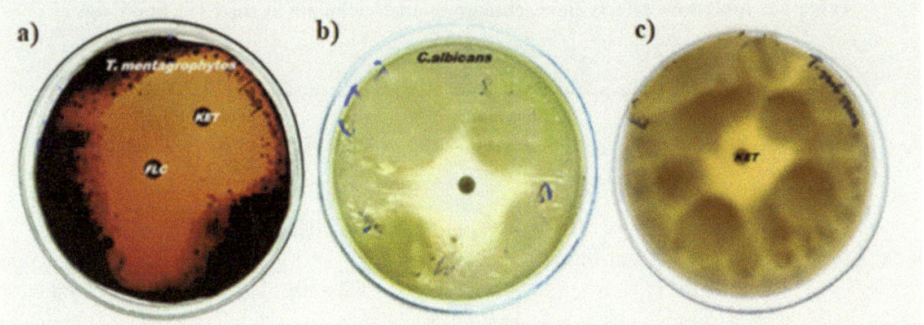

Figure 4.1. a) Shows zones of inhibition produced by standard drug Flc and Ket against *T. mentagrophytes*. **b)** Shows zones of inhibition produced by standard drug Flc against *C. albicans*. **c)** Shows zones of inhibition produced by standard drug Ket against *T. rubrum*.

4.2 Determination of antifungal effects of clove, cinnamon, kacholam against *T. mentagrophytes*

The result showed that clove exhibits significant fungicidal activity against *T. mentagrophytes. The* least concentration of clove (6.25mg) remarkably inhibits the mycelial growth of T. mentagrophytes. The standard drug *Ket* exhibit 27mm zone of inhibition. This fungicidal effect is more than that of standard *Ket* produced. Here also least concentration of kacholam (6.25mg) showed fungistatic activity against *T. mentagrophytes*. Least concentration showed 28mm zone of inhibition. This fungistatic effect is more than that of standard Ket produced. Highest concentration of kacholam (25mg) exhibit 35mm zone of inhibition against *T. mentagrophytes*. While the lowest 3 concentration of cinnamon showed moderate antifungal activity against *T. mentagrophytes*. But when the concentration of cinnamon is increased into 25mg, the antifungal activity also gets increased. Cinnamon also exhibit 33mm zone of

inhibition at 25mg concentration (*Figure 4.2 ii. a, b &c*). This means clove, kacholam, and cinnamon show more anti-fungal effect than standard drug ketoconazole. All the spice extracts tested considerably possess mycelial inhibition along with fungistatic activity. The results are shown in the table 4.2 and Figure 4.2i.

Table 4.2. Inhibitory effects clove, cinnamon and kacholam against Trichophyton
mentagrophytes

Spices	Zone of Inhibition (mm)				
	Antibiotics	Spices at different concentration			
		6.25mg	12.5mg	18.75mg	25mg
Cinnamon	27(K)	17	20	26	33
Clove	27(K)	S	S	S	S
Kacholam	27(K)	28	30	33	35

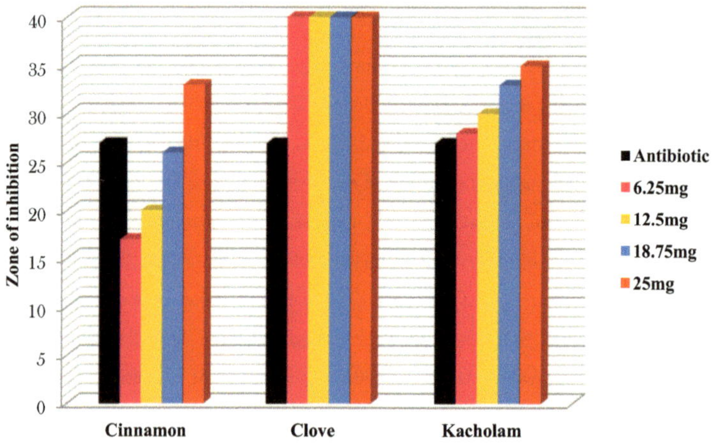

Figure 4.2. i) Graphical representation of inhibitory effects of clove, cinnamon and kacholam
against Trichophyton mentagrophytes

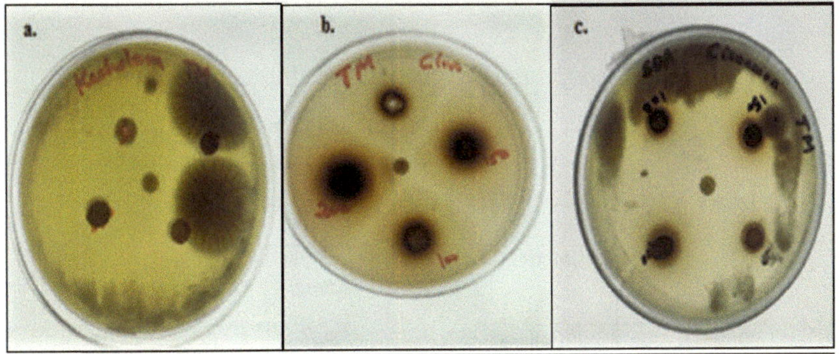

Figure 4.2. ii) a. Anti-fungal effect of kacholam against *T. mentagrophytes*. **b**. Anti-fungal effect of clove against *T. mentagrophytes*. **c**. Anti-fungal effect of cinnamon against *T. mentagrophytes*.

4.3 Determination of antifungal effects of clove, cinnamon, kacholam against *T. rubrum*

Table 4.3 and graphical representation (Figure 4.3 i) describes the inhibitory effects of clove, cinnamon, kacholam against T. rubrum. Results revealed that clove shows significant anti-fungal effect against *Trichophyton rubrum* with zone diameter of 39mm (25mg). Mycelial inhibition of *T. rubrum* by clove was well found in a concentration of 18.75mg and 25mg.Whereas standard drug Ket exhibit only 24mm zone of inhibition. Kacholam also exhibit good range of anti-fungal effect against *T. rubrum* with zone diameter of 23mm at higher concentration 25mg. Clove and kacholam showed fungistatic activity against *T. rubrum*. Cinnamon does not produce any inhibitory effect against *T. rubrum*. Results were shown in figure 4.3 ii.

Table 4.3. Inhibitory effects clove, cinnamon and kacholam against *T. rubrum*

Spices	Zone of Inhibition (mm)				
	Antibiotics	Spices at different concentration			
		6.25mg	12.5mg	18.75mg	25mg
Cinnamon	24(K)	R	R	R	R
Clove	24(K)	12	21	25	39
Kacholam	24(K)	10	16	17	23

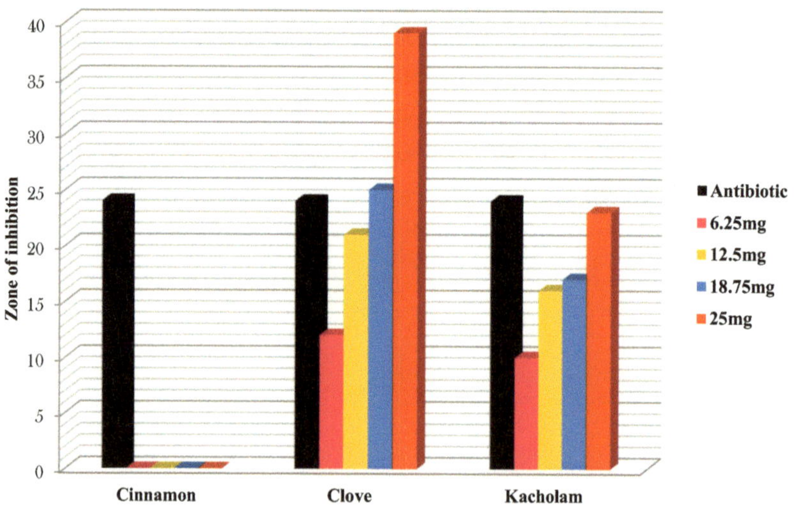

Figure 4.3. i) Graphical representation of inhibitory effects clove, cinnamon and kacholam against *T. rubrum*

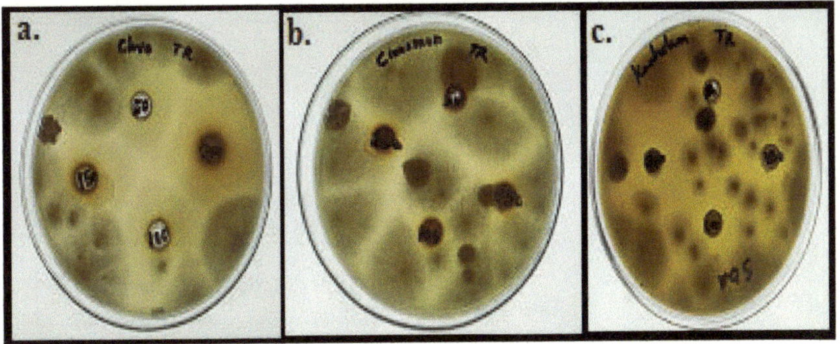

Figure 4.3. ii) a. Anti-Fungal effect of clove against *T. rubrum.* **b.** Anti-Fungal effect of cinnamon against *T. rubrum.* **c.** Anti-Fungal effect of kacholam against *T. rubrum*

4.4 Determination of antifungal effects of clove, cinnamon, kacholam against *C. albicans*

The present study revealed that Flc sensitive *C. albicans* strain showed sensitivity against both clove and kacholam. The graphical analysis (Figure 4.4 i) of *Table 4.3* describes the lowest concentration of clove extracts (6.25mg) exhibits 22mm zone inhibition against *C. albicans,* although standard drug Flc showed 22mm of inhibition zone. Highest concentration clove (25mg) showed 33mm zone of inhibition against *C. albicans.* This effect is more than that of standard produced. Cinnamon shows intermediate inhibitory activity against *C. albicans.* At higher concentration of cinnamon extracts showed 20mm zone of inhibition against *C. albicans.* Kacholam does not produce any effect against C. albicans (Figure 4.4 ii).

Table 4.4. Inhibitory effects clove, cinnamon and kacholam against *C. albicans*

Spices	Zone of Inhibition (mm)				
	Antibiotics	Spices at different concentration			
		6.25mg	12.5mg	18.75mg	25mg
Cinnamon	22(F)	12	13	18	20
Clove	22(F)	22	26	28	33
Kacholam	22(F)	R	R	R	R

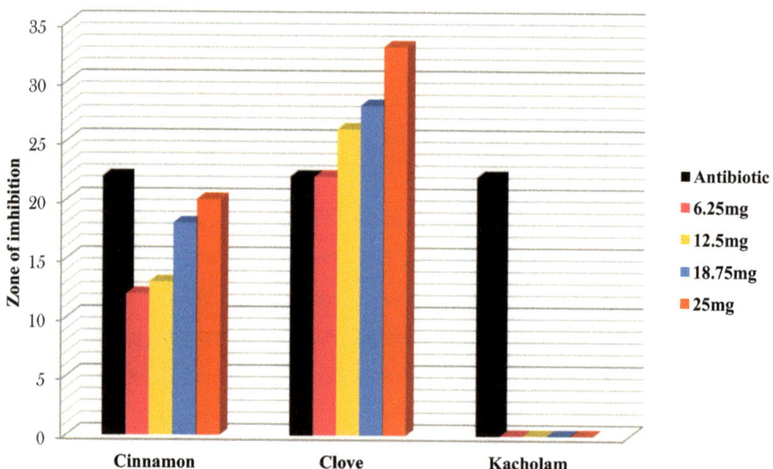

Figure 4.4. i) Graphical representation of inhibitory effects clove, cinnamon and kacholam against *C. albicans*

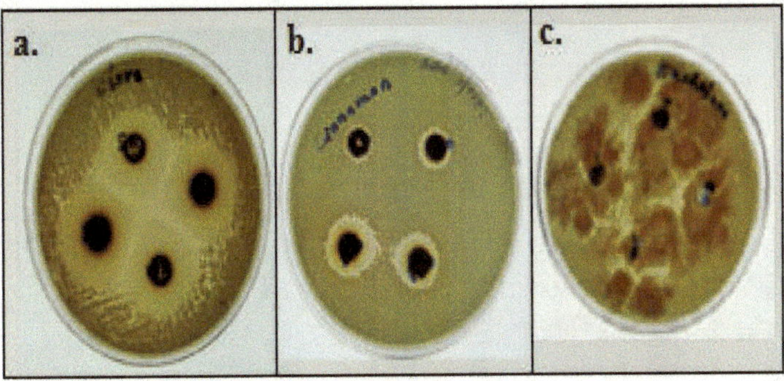

Figure 4.4. ii) a. Anti-fungal effect of clove against *C. albicans.* **b.** Anti-Fungal effect of cinnamon against *C. albicans.* **c.** Anti-Fungal effect of kacholam against *C. albicans.*

4.5 Determination of combinatorial effects of clove, cinnamon, kacholam against pathogenic fungi

All spices extracts were active against selected pathogenic organisms. But in combination the activity decreases as compared to that of individual effect. Table 4.5 and graphical representation (Figure 4.5 i) describes the combinatorial effects of clove, cinnamon, kacholam against selected pathogenic fungi such as *T. rubrum, T. mentagrophytes and C. albicans.* exhibits resistance against the combination of extract. The combined effect of spice extracts considerably inhibits *T. rubrum and T. mentagrophytes.* 25mg concentration of spice extract inhibits *T. rubrum* with a zone diameter of 33mm and *T. mentagrophytes with a zone diameter of 20mm.* Whereas standard drug Ket exhibit only 24mm zone of inhibition against *T. rubrum* and 27mm zone of inhibition against *T. mentagrophytes.* Results showed that the combinatorial effect of spice extracts does not produce any fungicidal or fungistatic activity against *C. albicans* (Figure 4.5 ii).

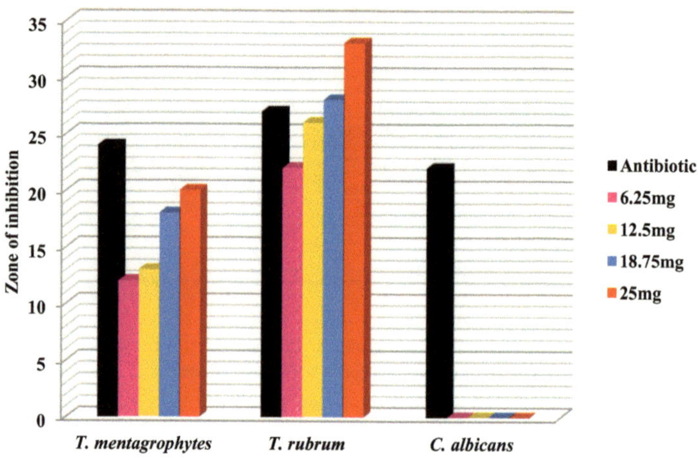

Figure 4.5. i) Graphical representation of inhibitory effects of combined extracts against pathogenic fungi

Organisms	Antibiotics	Zone of Inhibition (mm)			
		Concentration of spice combinations			
		6.25mg	12.5mg	18.75mg	25mg
T. rubrum	24(K)	22	26	28	33
T. mentagrophytes	27(K)	12	13	18	20
C. albicans	22(F)	R	R	R	R

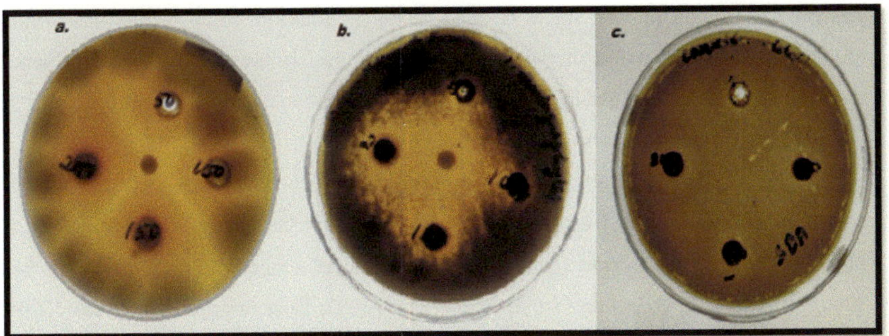

Figure 4.5. ii) a. Anti-fungal effect of clove-cinnamon-kacholam against *T. rubrum*. **b.** Anti-fungal effect of clove-cinnamon-kacholam against *T. mentagrophytes*. **c.** Anti-fungal effect of clove-cinnamon-kacholam against *C. albicans*

5. Discussion

Mucosal epithelial layer of human body is considered as nutrient rich niche of varies types of microorganisms, mainly bacteria and fungus. Survival of these microbial community is depending on both internal and external factors. A drastic change in these factors may leads to the breakdown of homeostasis, and thus it may lead to an infectious condition in the host (Albert et al, 1980; Bruce et al, 1973). The medical sector solved by these problems by the invention of antimicrobial drugs. Unfortunately, the microbial community develops drug resistance by changing their genomic structure by scale to point mutation. Development of drug resistance is the biggest problem of health care system during the 21st century (Andrea et al, 2019). Therefore, researches are focusing on natural plant based drug therapy for the treatment of microbial infection.

The current scientific research has revealed the surprising facts behind the medicinal properties of spices and defined their phytochemicals constituents which contributes its health benefits for pharmacological activities. Therefore, the current study aimed for the determination of individual and combinatorial effects of *Cinnamon, Clove,* and *Kacholam* against *T. rubrum, T. mentagrophytes* and *C. albicans*.

Results from the present study showed that each of the spice extract tested indicated a good antifungal activity on not less than two of the tested organisms. These differences in the antifungal action might be because of the differences in the chemical constituents present in each spices (Cowan, 1999). In addition, some factors like origin, climatic conditions in which they grow or the solvent used in the extraction will influence the activity of a natural product because all the different constituents present in it is dependent upon these factors (Ncube et al, 2008).

In this study the extract of clove significantly inhibits the mycelial growth of *T. rubrum* and *T. mentagrophytes* and also it exhibits fungistatic effects against *C. albicans*. These observations were in agreement with the previous studies, suggests that clove extracts had strong antimicrobial properties (Smith–Palmer et al, 1998; Delaquis et al, 2002; Burt, 2004; Arora & Kaur, 1999). Although the information regarding the mode of action is limited but the major components which are responsible for the antifungal actions of clove are eugenol, flavones, phenolic monoterpenoid glycosides, thymol, terpenes and aliphatic alcohols among other components (Deans&Ritchie, 1987; Dorman& Deans, 2000; Hammer et al, 1999). Furthermore, results of the study are in agreement with the study conducted by Rana et al. (2011), who evaluated the antifungal activity of clove oil against *T. rubrum*, reported that all the tested fungal species were inhibited by this oil.

The antifungal screening of Kacholam showed a good range its activity against *T. rubrum and T. mentagrophytes*. It inhibits the mycelial growth of *T.mentagrophytes* and thereby prevents the fungal growth. When the concentration of extracts is increased the fungistatic activity also gets increased. The rhizome of kacholam is rich in volatile oil, numerous alkaloids, starch, protein, aminoacids, minerals and fatty matter, of which Cinnamate as well as p-methoxycinnamate were the chief components in the rhizome oil of Kacholam. These compounds are thought to be the reason for the antifungal activity of kacholam (Arambewela et al, 2000; Ravindran et al, 2005; Wong et al, 1992; Ibrahim and Sadikun, 1990).

In the case of cinnamon, the present study proved its excellent inhibitory action against *T. mentagrophytes*. In contrast, the extracts of cinnamon also showed moderate antifungal activity against *C. albicans*. The results were in agreement with the study conducted by Cheng et al. (2009), suggested that cinnamon can act as antimycotic agent against certain fungi. The antifungal activity of cinnamon may be due to the presence of cinnamaldehyde, which prevents amino acid decarboxylase activity. Cinnamaldehyde, is a phytochemical found in the bark of cinnamon, which plays an important role in preventing microbial growth by interfering the electron transport chain and it reacts with nitrogen containing substrate (Gupta et al, 2008). Cinnamon act as a reservoir of numerous essential oils including b-caryophyllene, L-bornyl acetate, E-nerolidol, α-cubebene, α-terpineol *trans*-cinnamaldehyde, cinnamyl acetate, eugenol, L-borneol, caryophyllene oxide, terpinolene, and α-thujene (Gende et al, 2008; Wang et al, 2005; Prabuseenivasan et al, 2006). Therefore, these components will contribute to its antifungal action.

Although each spice extract showed difference in its inhibitory effect against different pathogenic fungi. However, the combination of clove, cinnamon and kacholam inhibits the growth of *T. rubrum* and *T. mentagrophytes* but C. albicans exhibits strong resistance against it. Therefore, the result reveals that the combination of spice extracts has antagonistic effect against *C. albicans* and it has only poor antifungal activity against *T. rubrum* and *T. mentagrophytes* as compared to that of its individual effect. Hence, it is clear that these combinations will not exhibit any synergism or additive effect to enhance the antifungal activity against the tested fungi.

6. Summary and Conclusion

Based on the observation of antifungal activities of clove, cinnamon and kacholam, it can be concluded that each spices have specific antifungal activity against each pathogenic fungus individually but combinations of these extracts do not exhibit potent antifungal activity. Traditionally spices are commonly used as food preservatives, flavouring agent and antiseptics, but it is important to study its antifungal action in order to explore their therapeutical applications. Further studies are necessary to examine the specific role of phytochemicals and its mode of action on growth inhibition. So that, these spices can be effectively use instead of conventional antifungal drugs.

> - *Clove was the most active and effective spice extract among all and it exhibits both fungicidal and fungistatic effect against T. rubrum, T. mentagrophytes and C. albicans.*
> - *Cinnamon and Kacholam exhibits moderate fungistatic effect against selected pathogens.*
> - *T. rubrum showed complete resistance against cinnamon extracts and C. albicans exhibits resistance against kacholam extracts.*

Figure 6.1. Schematic representation of antifungal effects of clove, cinnamon and kacholam against *Trichophyton rubrum*.[5]

Schematic representation 6.1 describes clove and kacholam possess fungistatic effect against T. rubrum whereas cinnamon showed no effect against T. rubrum.

---

[5] Author's own creation

Figure 6.2. Schematic representation of antifungal effects of clove, cinnamon and kacholam against T. *mentagrophytes*.[6]

Figure 6.2 describes clove extract showed significant fungicidal effect against T. mentagrophytes whereas Cinnamon and kacholam showed fungistatic effect or they prevent the mycelial growth of T. mentagrophytes.

---

[6] Author's own creation

Figure 6.3. Schematic representation of antifungal effects of clove, cinnamon and kacholam against *Candida albicans*.[7]

Figure 6.3 explains the clove inhibits C. albicans significantly than cinnamon. Cinnamon showed narrow range of antifungal effect against C. albicans while kacholam extracts does not produce any antifungal effect against C. albicans.

---

[7] Author's own creation

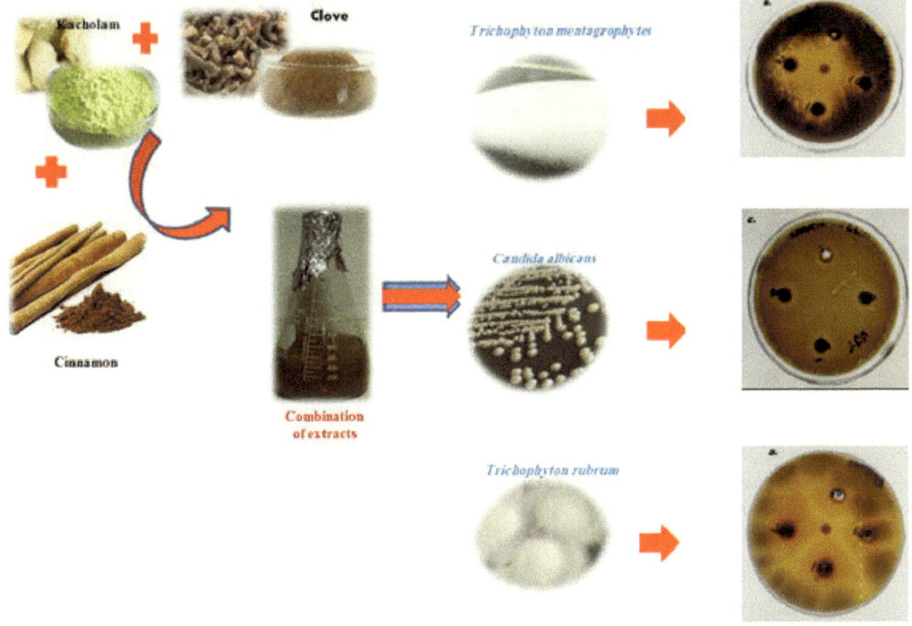

Figure 6.4. Schematic representation of combined effect of clove, cinnamon and kacholam against selected pathogenic fungi.[8]

Combinatorial antifungal activity of spice extracts was decreases as compared to that of individual effect. Combinatorial effect of clove, cinnamon and kacholam inhibits T. rubrum, and T. mentagrophytes but it has no effect on C. albicans.

---

[8] Author's own creation

References

Albert MJ, Mathan VI, Baker SJ, Vitamin B12 synthesis by human small intestinal bacteria, Nature, 1980; 283: 781–782.

Andrea MC, Rafael AA, Younes S, Drug Repurposing for the Treatment of Bacterial and Fungal Infections, Frontiers in microbiology, 2019; 10: 41.

Arambewela L, Perera A, Thambugala R, et al, Investigations on Kaempferia galanga, Journal of the National Science Foundation of Sri Lanka, 2000; 28: 225-230.

Arora D, Kaur J, Antimicrobial activity of spices, International Journal of Antimicrobial Agents, 1999; 12: 257-262.

Baratta MT, Dorman HJD, Deams SG, et al, Antimicrobial and antioxidant properties of some commercial essential oils, Flavour and Fragrance Journal, 1998; 13: 235–244.

Borelli D, Microsporum racemosum nova species, Acta Medica, 1965; 12: 148–151.

Bruce AW, Chadwick P, Hassan A, VanCott GF, Recurrent urethritis in women, Canadian Medical Association Journal,1973; 108: 973–976.

Burt S, Essential oils: Their antibacterial properties and potential applications in foods- a review, International Journal of Food Microbiology, 2004; 94: 223–253.

Chaieb K, Zmantar T, Ksouri R, et al, Antioxidant properties of essential oil of Eugenia caryophyllata and its antifungal activity against a large number of clinical Candida species, Mycoses, 2007; 50: 403-406

Chang ST, Chen PF, Chang SC, Antibacterial activity of leaf essential oils and their constituents from *Cinnamomum osmophloeum*, Journal of Ethnopharmacology, 2001;77: 123–127.

Cheng SS, Liu JY, Huang CG, et al, Insecticidal activities of leaf essential oils from *Cinnamomum osmophloeum* against three mosquito species, Bioresource Technology, 2009; 100: 457–464.

Conti HR, Huppler AR, Whibley N, Gaffen SL, Animal models for candidiasis, Current Protocols in Microbiology, 2014, 105: 19.6.1–19.6.17.

Cortes-Rojas DF, De Souza CRF, Pereira OW, Clove (Syzygium aromaticum): a precious spice, Asian Pacific Journal of Tropical Biomedicine, 2014; 4: 90-96.

Cowan MM, Plant products as antimicrobial agents, Clinical Microbiology Review, 1999; 12: 564–582.

Deans SG, Ritchie GA, Antimicrobial properties of plant essential oils, International of Journal Food Microbiology, 1987; 5: 165- 180.

Delaquis PJ, Stanich K, Girard B, Mazza G, Antimicrobial activity of individual and mixed fractions of dill, cilantro, coriander and eucalyptus essential oils, International Journal *of* Food Microbiology, 2002;74:101–109.

Di Paoli S, Giani TS, Presta GA, et al, Effects of Clove (Caryophyllus aromaticus L.) on the Labeling of Blood Constituents with Technetium-99m and on the Morphology of Red Blood Cells, Brazilian Archives of Biology and Technology, 2007; 50: 175-182.

Dorman HGD, Deans SG, Antimicrobial Agents from Plants: Antibacterial Activity of Plant Volatile Oils, Journal of Applied Microbiology, 2000; 88: 308-316.

Fuentes CA, A new species of Microsporum, Mycologia, 1956; 48:613–614.

Gende LB, Floris I, Fritz R, et al, Antimicrobial activity of cinnamon (Cinnamomum zeylanicum) essential oil and its main components against paenibacillus larvae from argentine, Bulletin of Insectology, 2008; 61: 1–4.

Georg LK, Ajello L, Friedman, et al, A new species of Microsporum pathogenic to man and animals, Sabouraudia, 1962; 1:189–196.

Gupta C, Garg AP, Uniyal RC, et al, Comparative analysis of the antimicrobial activity of cinnamon oil and cinnamon extract on some food-borne microbes, African Journal of Microbiology Research, 2008; 2:247-51.

Hainer BL, Dermtophyte infections, American Family Physician, 2003;67:101-8.

Hainer BL. Dermtophyte infections. Am Fam Physician.

Hainer BL. Dermtophyte infections. Am Fam Physician.

Hammer KA, Carson CF, Riley TV, Antimicrobial Activity of Essential Oils and Other Plant Extracts, Journal of Applied Microbiology, 1999; 86: 985-990.

Ibrahim P, Sadikun A, Antimicrobial studies on ethyl pmethoxycinnamate and ethyl cinnamate from Kaempferia galanga. Proceedings of the 7th National Seminar on Natural Products, Universiti Sains Malaysia, Penang, 1990: 174–178.

Irene W, Richard C, Summerbell, The Dermatophytes, American Society for Microbiology, 1995;8: 240–259.

Jakhetia V, Patel R, Khatri P, et al, Cinnamon: A pharmacological Review, Journal of Advanced Scientific Research, 2010; 1: 79-23.

Kashem SW, Igyarto BZ, Gerami-Nejad M, et al, Candida albicans morphology and dendritic cell subsets determine T helper cell differentiation, Immunity, 2015; 42: 356–366.

Kohner EM, Stratton IM, Aldington SJ, et al, Microaneurysms in the development of diabetic retinopathy (UKPDS 42), UK Prospective Diabetes Study Group, Diabetologia, 1994; 42: 1107–111.

Langner E, Greifenberg S, Gruenwald J, Ginger: History and use, Advances in Therapy, 1998; 15:25–44.

Lucy H, Edgar JD, Medicinal plants: a re-emerging health aid, Electronic Journal of Biotechnology, 1999; 2: 56-70.

Martins MD, Lozano-Chiu M, Rex JH, Point prevalence of oropharyngeal carriage of fluconazole-resistant Candida in human immunodeficiency virus-infected patients, Clinical Infectious Diseases, 1997; 25: 843-846.

Martins N, Ferreira IC, Barros L, et al, Candidiasis: predisposing factors, prevention, diagnosis and alternative treatment, Mycopathologia, 2014; 177: 223-40.

Mathabe MC, Nikolova RV, Lall N, et al, Antibacterial Activities of Medicinal Plants Used for the Treatment of Diarrhoea in Limpopo Province, South Africa, Journal of Ethnopharmacology, 2006;105: 286-93.

Minich St, Msom L, Chinese Herbal Medicine in Women's Health, Women's Health; 2008.

Murray PR, Baron EJ, Pfaller MA, et al, Manual of Clinical Microbiology, 6th Ed, Washington DC, ASM Press; 1995.

Ncube NS, Afolayan AJ, Okoh AI. Assessment techniques of antimicrobial properties of natural compounds of plant origin: current methods and future trends. Africa Journal of Biotechnology, 2008; 7: 1797–1806.

Parle M, Khanna D, Clove: A champion spice, International Journal of Research in Ayurveda and Pharmacy, 2011; 2: 47-54.

Prabuseenivasan S, Jayakumar M, Ignacimuthu S, In vitro antibacterial activity of some plant essential oils, BMC Complementary and Alternative Medicine, 2006; 6.

Rana SI, Rana SA, Rajak CR, Evaluation of antifungal activity in essential oil of the Syzygium aromaticum (l.) by extraction, purification and analysis of its main component eugenol, Brazilian Journal of Microbiology, 2011; 42: 1269-1277.

Ravindran PN, Balachandran I, Underutilized medicinal spices (II), Galanga (Kaempferia galanga L.), Spice India, 2005; 18: 22-35.

Rex JH, Rinaldi MG, Pfaller MA, Resistance of Candida species to fluconazole, Antimicrobial Agents and Chemotherapy, 1995; 39: 1-8.

Sarker SD, Nahar L, Chemistry for Pharmacy Students General, Organic and Natural Product Chemistry, John Wiley and Sons, England, 2007: 283-359.

Smith–Palmer A, Stewart J, Fyee L, Antimicrobial properties of plant essential oils and essences against five important food borne pathogens, Letters in Applied Microbiology,1998; 26: 118–122.

Wang SY, Chen PF, Chang ST, Antifungal activities of essential oils and their constituents from indigenous cinnamon (Cinnamomum osmophloeum) leaves against wood decay fungi, Bioresource Technology, 2005; 96: 813–818.

Wondrak GT, Villeneuve NF, Lamore SD, et al, The cinnamon-derived dietary factor cinnamic aldehyde activates the Nrf2-dependent antioxidant response in human epithelial colon cells, Molecules, 2010;15: 3338–3355.

Wong KC, Qng KS, Lim CL, Composition of the essential oil of rhizome of Kaempferia galanga L, Journal of Flavour and Fragrance, 1992; 7: 263-266.

YOUR KNOWLEDGE HAS VALUE

- We will publish your bachelor's and
  master's thesis, essays and papers

- Your own eBook and book -
  sold worldwide in all relevant shops

- Earn money with each sale

Upload your text at www.GRIN.com
and publish for free